I0488862

Cash in the Wind:

How to Build a Wind Farm using Skystream and 442SR Wind Turbines for Home Power Energy Net-Metering and Sell Electricity Back to the Grid

by Christopher Kinkaid

 Solardyne.com

Published by Solardyne, LLC
Portland, Oregon

ISBN-13: 978-1500483807
ISBN-10: 150048380X

Table of Contents

Preface

The power of wind is enormous. Tap into this incredible power supply, using state of the art wind turbines, to generate electricity for sale to the Grid. Wind power, worldwide, has been the fastest growing installed clean energy power supply. Now, you can Harvest your Wind Energy for Profit.

How can you harvest this gold? How can you Cash-in the Wind?

This Book describes how to Build a Wind Farm, using Skystream and 442SR Wind turbines, to "mine" wind energy on your property safely, properly, and profitably. The wind industry has "evolved" over the last 30 years, and has emerged as a world-class industry, with remarkable growth. Wind Turbine Hardware has matured offering the industry reliability, safety, and long life in the field.

Major utilities, and Independent Power Producers, have tapped into Large Wind Farms with Megawatt output. This Book is written to assist in Small Wind Farms, suitable for your Home, Farm, Ranch, Business, and Commercial power needs from 500 to 20,000 kWh per month.

Wind Farms, historically, been only the realm of Large Independent Power Producers. Now, with Today's high quality equipment, Wind Turbines are available to anyone with an electric bill, and a suitable wind resource.

This Book covers Small Wind Farms for Residential, Commercial, and Light Industrial power consumers, or Independent Power Producers. Wind Farms, discussed in this Book, range from Monthly Energy Production from 500 to 20,000 Kilowatt-hours.

About the Book

This Book is written as a step-by-step guide to defining your Wind Farm "vital statistics," and choosing the right equipment to get the job done. If you have a specific Power Rating, or KWH Energy rating in mind for your Wind Farm, then see Wind Farm System Examples located at the **Quick Guide** in Chapter Nine.

The **Quick Guide** contains Example Systems which take you to specific Wind Farm layouts designed to generate specific Energy Monthly Outputs. Check your Electric Bill and find your Energy kWh consumption per month. The Quick Guide also lists links to Wind Farms systems by Power Rating, used for Rebate Calculations.

Match your Monthly kWh consumption with the system on the list which most closely matches your load. If you're building a new Wind Power system to produce electricity under a Power Purchase Agreement, choose your System by the closest Monthly kWh rating you wish to produce. Wind Energy Systems range from 500 kWh to 20,000 kWh energy output per month.

In **Chapter 1,** looks at Wind Power in the big picture, with overview of approaching a Wind energy generating facility. Looks at the Power in the Wind, and how Wind Resources are categorized.

Chapter 2, describes how to Define your Wind Power Facility project.

Chapter 3, the Skystream 3.7 Wind Turbine for Home Power Grid Tie use.

Chapter 4, reviews different Skystream 3.7 Configurations for Monthly energy production from 500 to 2,500 kWh per month.

Chapter 5, looks at the larger 442SR Wind Turbine rated at 10,000 watts for use in Wind Farms.

Chapter 6, reviews Wind Power system examples for multiple 442SR Wind Turbines from 2,500 to 20,000 kWh monthly energy production.

Chapter 7, Installing your Wind Farm examines the issues of Site Selection, Site preparation, Foundation, and Tower aspects of your Wind Project.

Chapter 8, Opportunities in Wind Farms reviews Net-Metering, and Power Purchase Agreement projects

Chapter 9, Quick Guide to Wind Energy Systems from 500 to 20,000 kWh Monthly Energy Production.

About the Author

Christopher Kinkaid

Christopher (Toby) Kinkaid, originally from Portland, Oregon is the founder of **Solardyne.com**, **SolarQuote.com**, and **AlgaeToday.com**, and has worked in clean energy technology for over three decades.

Kinkaid, is the inventor of the "**Helyx**"Vertical Axis Wind Generator, the "**Mariposa**" Non-imaging solar concentrator PV module (continuous operation at Sandia National Laboratory since 1994), the **Solar Demultiplexer** optical solar concentrating lens (Dr. James/Sandia National Laboratory 1991), and the inventor of the original "Solar Power Pack" (Mother Earth News, "Littlest Utility" June/July, 2001).

Kinkaid, has been an official lecturer and presenter on clean energy technology around the world including APEC, Bangkok, Thailand, 2003, "Energy Solutions World", Tokyo, Japan, 2003, The International Biomass Conference (IBC), 2010,

Minneapolis, MN, and the Algal Biomass Organization (ABO) Conference, 2010, Phoenix, AZ.

Kinkaid, has appeared in interviews on KOIN TV, KGW TV, and "Sustainable Today" produced in Oregon. Kinkaid, has served on the board of directors for the National Hydrogen Association, in Washington D.C., 1993, and the Japan Satellite Communications Company (JCNET), Fukuoka, Japan, 1994-95.

Kinkaid, served on the board of directors for Algaedyne Corporation, Preston, MN, 2010-2013. Kinkaid, presently serves as CEO of Solardyne, LLC in Portland, Oregon.

Christopher Kinkaid is based on the West Coast, and continues his work in Solar, Wind, and Biomass applications, research, and development in Portland, Oregon.

Introduction

There's a fortune in wind blowing across your land. If Gold were flying through the air, would you reach out and scoop some up? Wind power, is another form of gold, and modern Wind Turbines give you an effective means of producing power, (energy) on your land, for sale to your Utility.

Wind turbines, convert your Wind into cash by converting physical power into electricity for export to your Grid. This Book covers Two models of Wind Turbines designed for the residential, and commercial markets. Wind turbines can be used, singularly, or combined into a "mini" Wind Farms to increase energy production.

This Book examines the essential elements of a Wind Farm, using the Skystream, and 442SR Wind Turbines, then suggests some System configurations for various Power, and Energy outputs.

Wind power, and Human activities go way back. Some of the earliest "industrialization" were ancient Vertical Axis wind turbines fashioned on vertical poles, with cross members "webbed" with stretched hides. Shallow vertical holes in the ground, with clay pots at the bottom, served as "hubs" and provided a great natural bearing.

As the wind blowed, the Contoured stretched hides caught the breeze, and would rotate. Neolithic humans poured grain down the holes into the buried pots, and the turning shaft would grind the grain. Evidence of these techniques date back to Neolithic Turkey, and China.

Amazingly, where "winds" were prevailing, the Neolithic cultures would construct "walls" which "shelter" the up-wind side of the vertical turbine. "Blocking" the wind on the Up-stream side, improved performance, and power to grind the grain.

Moving winds, like moving water, were surprisingly effective as a power supply in ancient times. Fast forward, to early "Dutch" wind mills, where Vertical Axis were replaced by Horizontal Axis architecture, and power increased again. These massive wind machines, with vanes catching the wind turning shafts "horizontal" to the ground, produced enormous torque. The enormous force of the wind, captured, and converted with these Grain mills were made of stone, were readily rotated by the enormous torques produced with vanes stretching dozens of feet from the central hub.

Wind power, for effective commercial use, predates the Industrial Age, with Megawatts of installed capacity in Europe before the Middle Ages. An impressive feat, and a statement to the "practicality" of wind power. Today, modern wind turbines stand on decades of improvements, and have reached a state-of-the-art to be reliable, safe to operate, and productive in the field.

Chapter One - Wind Power the Big Picture

In this book, we'll break down the questions you'll need to ask to define your system requirements leading to a Wind Farm solution. Let's take a look at the nature of Wind, and how it effects the production of electricity.

The wind is a powerful, and ever changing, natural resource. Wind generators operate differently than "normal" generators you power with fuel. A diesel Generator set (Genset), for example, burns fuel to run the Generator at a particular output, for a particular time length (limited by the size of your tank), to power your load.

The Diesel Genset delivers this reliability, but of course, has fuel-costs, as well as Capital, and

Mechanical costs. Gensets require planned maintenance, and usually must be replaced every 5 years depending on hours used.

The Wind turbine has no fuel cost. Robust, and long lived with planned maintenance your Wind Turbine is engineered for 20 year operational life.

Wind turbines, have these advantages, but operate differently than "normal" fuel-based generators. A wind turbine must deal with a Constantly changing resource, under ever changing conditions.

The wind changes in Speed, Direction, and turbulence many times each second. The physical stresses on wind generators are enormous, not just for the force of the wind, but how quickly those forces change.

Wind turbines, are designed to function in the real world. The world of many changing forces, stresses, and extremes while generating electricity with efficiency.

Wind turbines endure temperature swings, high speed particulates in the air, chemical corrosive chemicals, Salt water, and host of other "environmental" conditions including vibration, lightning, and vandalism. Further, a modern wind turbine must deal with the raw power in the wind, capture, and convert those forces into rotational torque to drive the generator.

Wind turbines are engineered, and made robust to deal with the powerful forces of the wind. A goldmine if you can catch it, convert it, and sell it.

The Power in the Wind

The power in wind is an awesome force all over the Earth. The wind is a fluid. Fluid-dynamics defines the study of fluids, for wind turbines the field is aerodynamics. As a special case of fluid-dynamics, moving air, (wind), aerodynamics follow the same physics.

The power in a moving fluid is a function of its density, considered cross-section area, and "the Cube" of the speed. The formula for power in a moving fluid is P = 1/2 (density of the fluid) times (Cross-section area considered) times the Velocity Cubed.

If you double the wind speed, the Power increases 8 Fold! (2 cubed). Moving fluids have 800% more power when moving twice as fast. Incredible.

Tornados, and Hurricane winds can destroy buildings in seconds, as power "exponentially" increases with wind speed. The power in the wind is an incredible resource - and one you tap, and Harvest, with your Wind turbines. Proper hardware choices, with proper installation and maintenance is the key.

Moving fluids, water and wind, increase in Power "exponentially" in an extreme way when moving faster with the Cube Rule. The power in the wind increases 64 fold from 7 mph to 28 mph. Wind Zone One power density is 200 watts per square meter. Wind Zone Seven has a power density of 2,000 watts per square meter. (See Wind Zones, below).

Your Site's Wind Resource

Wind turbines convert the physical force of the wind into Hard Currency: kWh of energy. Energy has industrial value everywhere. Producing energy for export for hard currency income goes to the economic performance of your wind power plant.

The "Wind Resource" is quantified by assigning every location on Earth an "average wind speed" in Wind Zones.

The "Resource" of wind is defined, specifically as it's power density (Watts per square meter) , and is "Categorized" into Seven (7) Wind Zones.

Zone One (I) is the lowest power Wind Zone. Zone 7, (VII) is the highest Wind Zone, and has the highest average wind speeds, and thus Power.

Your prospective Wind Farm site will have a **Wind Zone Number** from One to Seven (Zone I-VII) assigned. **Find your Wind Zone** on the Wind Map for your Site. Checking Wind Maps, will tell your locations "Wind Zone" and resource. Data usually collected at 80 meters height.

Propeller based Wind Turbines work best in Wind Zone 2, (Zone II), and higher.

Wind Zone's are defined by the average "power density" within a given wind speed, per square meter.

Wind Zone One (I) average power density 200 watts/m2

Wind Zone Two (II) average power density 300 watts/m2

Wind Zone Three (III) average power density 400 watts/m2

Wind Zone Four (IV) average power density 500 watts/m2

Wind Zone Five (V) average power density 600 watts/m2

Wind Zone Six (VI) average power density 800 watts/m2

Wind Zone Seven (VII) average power density 2,000 watts/m2

Wind Zones power density goes from 200 to 2,000 watts from 4.4 m/s to 9.4 m/s, as the "Cube" rule dictates "exponential" power gain as wind speed increases.

Wind Zones are defined by the Average Wind Speed:

Wind Zone One (I) average Wind Speed 0-4.4 m/s

Wind Zone Two (II) average Wind Speed 4.4 - 5.1 m/s

Wind Zone Three (III) average Wind Speed 5.1 - 5.6 m/s

Wind Zone Four (IV) average Wind Speed 5.6 - 6 m/s

Wind Zone Five (V) average Wind Speed 6 - 6.4 m/s

Wind Zone Six (VI) average Wind Speed 6.4 - 7 m/s

Wind Zone Seven (VII) average Wind Speed 7 - 9.4 m/s

Every Location has particular "**micro-climate**" conditions. Obstacles, Buildings, Higher areas, trees, and foliage all impact wind speeds. You want the highest location for your wind site, and ideally, above surrounding land, or obstacles for a 250 foot radius.

Your proposed Wind Site's "wind resource" is a function of Wind Zone, and Tower Height.

Electrical Definition of your Wind Farm.

All wind turbines, from a single to multiple units in a Wind Farm are all defined Electrically. Building a Wind Farm begins with defining your system from an "Electrical" point of view. How much Energy per Month do you want to produce? This is really the key question. From your "Energy" objective will come all other details.

In planning your Wind Farm, let's begin at the beginning: How much Energy do you want to produce each Month?

If you're a home owner you can look at your electric bill for your monthly energy demand in kWh per month. The Energy Demand, would be the basis for designing your Net-Metering wind power system.

Your "interconnection" with the Utility Grid determines which Model of Wind Turbine you'll choose. If you're using Single Phase 120 VAC at your site, then you're Model will be matched to 120 VAC/ 60Hz. If you're using 3 Phase power, for compressors, refrigerators, etc., then your Wind Turbine Model must be matched to a 3 Phase output. Your Utility will tell you what Electrical Service your site uses, if this information is unavailable.

Site Selection

The power in your wind, at your location, depends on the physical layout of your geography. As wind power increases with speed, Speed increases with height. Wind turbine Towers, are an important consideration. The higher the tower, the faster the wind speed. Faster wind speeds generate higher electricity production.

Choose your site to be clear of obstructions which slow the wind, and causes turbulence. Choose your site with appropriate access for pickup trucks, in a tilt up tower, or Cranes, for a stacked tower.

Choose your site to be "close" to your electrical interconnection with the Grid. This may be a Junction Box, Power Box, Transformer, or other electrical interface with your site. Choose your site such that your "Set Back" is appropriate. Set Back is the distance from your Wind Tower Base to your property line.

Examine your soil type, to help determine the proper foundation specs. Your Wind Manufacturer will have this data.

Site Access

Choosing your site must consider access. Seems obvious, but requires review. The Skystream 3.7, and Xzeres 442 Wind Turbines, discussed in this

Book, can be installed with Pick-Up truck equipment and Cranes, depending on your Tower choice. Site Access is not only a construction, but also a maintenance, and inspection consideration.

Wildlife and Avian Impact access. Humans, are not the only life forms involved with power plants. All power plants have impacts. Wind Turbines present a hazard to "prey" birds. Site your wind turbines to avoid any known migratory paths, or near protected areas.

If your wind site is on Federal Lands, when you negotiate your "easement", or "variance" you'll need to check with the BLM regarding access road permits.

Hardware Selection

Energy production is the key to Wind Farm sizing. Once you decide the Monthly Energy production objective, the next consideration is your Site Selection. Your site selection will determine your "**Wind Resource**." Once you know your energy objective, for example, a home owner with 700 kWh per month consumption, you can determine your Wind Hardware Size.

Wind turbines are rated in both Power rating, usually at 25 mph wind speed, with Energy Ratings depending on **Wind Zone**, (Link to NREL wind power atlas), and Height of Tower.

Once, you know your Energy Production target, contact the Wind Manufacturer to confirm the Energy Output of a given wind turbine, for a given Tower Height, for a given Wind Zone location. The Quick Guide in this book is an easy way to estimate the Wind Turbines you'll require for given, general, output.

Wind Turbine Towers

Tower choices are very important for Wind Farm performance, and economics. Higher towers have higher costs. However, performance increases, because the height exposes the Wind Turbine to higher average wind speeds - producing more energy over time.

Towers must be pre-engineered. Manufactures must design towers for large mechanical, vibrational, chemical, and abrupt stresses. Professional Tower engineering is vital for safety, and long life for your Wind Farm. Towers are engineered for specific Wind Turbines Models, specific Tower Heights, and specific Soil-types, at your Wind Site.

Towers, for a given Wind Turbine, are either Lattice-type, or Monopole-type. Each type requires construction, and assembly onsite. Both types can either be Tilt-up, or Crane "stacked" on your Foundation Base Plate.

Chapter Two - Defining your Wind Farm

Modern wind turbines make tapping into your wind power easy, and supported. The focus of this eBook is on Wind Farms you can build using the Skystream 3.7, and Xzeres 442 Model wind turbines. Wind turbines can be configured to charge battery systems, or provide electricity directly to the Grid.

This eBook will focus on Grid-tie Wind Farms. Grid-Tie wind farms to sell electricity back to your utility.

There are two ways you can sell electricity to your Utility. Net-metering, or Power Purchase Agreements (PPA). Net metering is one approach, but must be approved by your Utility, with a Net-metering policy in place. The other option? Become an Independent Power Producer (IPP), and approach your utility requesting to negotiate a Power Purchase Agreement (PPA).

Utilities routinely use PPAs to contract with Independent power producers IPPs. Utilities source "Energy" from a variety of providers. These IPPs may be powered by coal, oil, natural-gas, solar, hydro, and wind.

Utilities, in modern times, own less of their own generators, and "contract" for IPPs to provide power to their grid. This is your opportunity to sell wind produced electricity to your Utility.

Net-metering is a Federally protected right under the PURPA act (Public Utility Resource Policy Act). However, only publicly owned utilities (POU) must comply. Investor owned utilities (IOU) are not required to comply, though usually have some kind of wind connection program. You must contact your Utility, and ask if they support "Net-metering" with Wind Turbines. Utilities will have a Pre-Approved List of Manufacturers, and Wind Turbine Models.

The Wind Turbines listed in this Book are UL Listed, and known by most utilities, and are on most qualified equipment lists.

Net-metering allows you to put electricity from your wind turbine back into the grid against your bill. There are no batteries involved. If the wind is blowing faster than 7 mph, then your wind turbines will kick-in, and begin to produce electricity, which "export" your electricity to the grid. Wind turbines for Net-Metering are best sized, Just Under your Monthly kWh demand.

Wind electricity, with Net metering, will give you "full" retail value for your kWh of wind energy produced. However, some utilities, use a Time dependent meter. In this case, the utility will give you different prices for your wind energy, depending on the time of day the energy is produced. Energy produced during Peak times, 10 am to 6 pm, or how your utility defines peak, are worth more than energy produced during off-peak times, such as at night.

Most Utilities are Not time dependent with Net-Metering regarding "Time of Day." Net-Metering is usually compiled at the end of the Month. Your Wind Power electricity "production" in kWh, is deducted from your energy "Demand." You only pay for the Net difference, this is the essence of Net-Metering.

As your wind turbines produce electricity, the "smart" inverter is monitoring the Grid Electricity power factor, and bridges the Voltage-Current between wind turbine, and grid power. By "leading" the power factor a relatively "small" amount of current can be added to the grid. If a generator "lags" the grid power factor, then it becomes a load, and the "generator" becomes a "motor."

Modern wind turbines have sophisticated "power conditioning" protocols which efficiently "push" your wind turbine produced electricity back into the grid. This turns your "Electric Meter" backwards, as such. At the end of the month, under Net-Metering, the Total amount of kWh you produce is subtracted from the Total amount of kWh you consumed. You only pay the utility the "Net" difference between the two.

Smart Electric Meters are either "Dual" function with two meters, one for energy in, the other energy out. The other type of Smart Meters are "Bi-Directional" allowing metering in real time with one meter.

Note: Utilities can offer you retail value for your wind turbine generated electricity, however, if you produce More energy than you consumed, utilities will only pay the "Avoided Cost." Avoided cost, can be a "very low" price, such as 1.5 cents/kWh for your energy, so be sure to check.

To get the highest value for your wind energy, for Net-Metering, size your wind facility output (energy rating) just (less than) your Monthly energy consumption.

This insures you'll get the highest monetary value (retail) for your wind energy. Use the Following Steps to define your proposed Wind Farm:

Step One: How Much Energy do I Need to Produce Each Month?

Let's take the example of a home owner. Look at your electric bill and find your Total kWh Consumption per Month. Let's say it's 700 kWhs (Kilowatt-hours) per month. If your system is a Grid tie Net-metering system

Step Two: What is my Wind Resource at my Proposed Site?

Your wind resource is how much "gold" you have flying across your land every month.

The wind resource at your Wind Farm Site tells you how much wind you can tap. There are two parts to your Wind Resource determination. First, the general Wind Zone Rating of your location. Second, the "relative" wind resource, as impacted by local conditions (obstacles), and at What Height the wind speed is measured.

Wind travels faster, generally, the higher the altitude. Ground effects, such as buildings, trees, rises, depressions, generally, slow down the wind. You want your Wind Turbines as high as possible to tap higher wind speeds.

Wind speeds are usually sampled at two heights: 20 meters, and 80 meters. Large wind turbine projects, using the Large Megawatt class Vestas Wind

Turbines, with very high towers, will take the wind speed samples at 80 meter height.

Smaller wind turbines, less than 100,000 watts, (the wind turbines described herein), usually use the 20 meter measurement.

Anemometers measure wind speed. Handheld, or placed up on your tower, or a sampling tower will give you "instantaneous" wind speed values, for accurate date, you'll need a long sampling rate. Professionals sample for months, and even years in some cases, but for our purposes, manufacturers have already run these tests.

Once, you choose the Wind Turbine, or Turbines for your project, and the height of your towers, the Manufacturer can give you good Energy Production estimates.

Step Three: Does my Site give me the Best Wind Access?

A rule of thumb follows that your Tower height is limited by what it hits if it falls down. A primitive rule, but highly effective. Set-back, is the distance from your Wind Turbine Tower base to your property line. Be sure to site your Turbines, ideally, at a "set-back" as high as your Wind Turbine Tower.

Choose your Wind Turbine site as the highest elevation. The Wind Turbine should be at least 25 feet above any obstruction within 250 feet of your turbine.

Note: Home Turbines are sometimes limited by code, or incentive rules, to be within 100 feet of your Junction Box. If you're not limited by code, set your Wind Turbine from 100 to 200 feet from your Home. Often, rural areas are not limited by the 100 foot convention.

Step Four: Does My Utility Support Net-Metering, or PPA?

How you sell electricity to your utility depends on their Net Metering, and Power Purchase Agreement (PPA) protocols. If you're a home, or business owner, with an existing "Electric Bill," you'll use your Utilities Net-Metering program. Net-metering will give you the "highest" value (cents/kWh) for your wind produced electricity.

If your planning a Wind Energy power plant to sell electricity you'll need to negotiate a Power Purchase Agreement as an Independent Power Producer (IPP).

Step Five: What Are Regulatory Affairs?

"Regulatory Affairs" deal with all aspects of your Wind project. All physical impacts of your Wind Farm facility, and all "reporting" you need to generate to satisfy permitting is included in Regulatory Affairs. In general, a preliminary document which describes your wind project would include an "Impact Study." Your Wind Facility description, in written form, can follow two documents. These two documents are your Site Plan, and your Impact Study.

Your site plan is a description of your Wind project from start to finish. Describe your Site Location's Wind Resource, the Electrical Energy Production, the Wind Turbines you've chosen to install, and commission, and your Installation Schedule with tasks. Your Impact Study, if required, will include Air Impacts, Soil Impacts, Water Impacts, Wildlife Impacts (Avian Impacts), Anthropological, and Paleontological impacts.

The good news, is Most of these impacts are minimal for smaller Wind Turbines, and an extensive write up is, often, not required. Regulatory Affairs is a topic you'll use, IF you're required to produce project documentation. Your Utility, or State Energy office will have forms you can use to fulfill this requirement if your Wind Project is a Net-metering facility.

If you're going to build a Wind Farm exclusively for energy export to a grid, using a Power Purchase Agreement (PPA), then regulatory affairs will be more stringent with paperwork regarding your Site Plan, and your Impact Study documents.

The first step is to contact your Utility. Ask whether they support Net-Metering, and if so, to direct you to their policy page. Your Electric Service Provider will have a program that manages Grid Interconnection from Renewable Sources, this includes Solar Photovoltaics, and Wind Turbines, such as the Skystream, and 442SR.

Step Six: Does my Wind Farm Quality for Rebates, Tax Credits, Grants, and Incentives?

The great advantage of building your own wind farm is clean energy production, and the Financial Incentives available to you - the owner.

Wind turbines are not cheap. You 'get what you pay for' with wind turbines. The reason? The real world is very tough on Wind Turbine hardware. Each Season presents its own regime of temperature, and physical stresses, and challenges for Hardware in the field. Engineering must be extremely robust.

The Wind turbines described herein are field-proven, UL-Listed, and already registered with Qualified Equipment Lists throughout the country for rebates, and financial incentives. Each State, and each Utility will have their own rebate, grant, and incentive program. Federally, your wind site qualifies for a 35% tax credit. The USDA offers a 25% rebate, once your wind turbine is installed, depending on whether your site is in a Rural zone.

See Chapter Eight for additional discussion about Rebates, Tax-Credits, Grants, and Incentives.

Step Seven: Configure your Wind Power System

We've considered aspects of your Wind Farm project, not let's define Specific Hardware. The "Power" rating of your wind turbine, or number of wind turbines, depends on your locations Wind Zone, and specifically, the height of your wind speed measurement. To Choose your Specific Wind Turbine, or Wind Turbines, to reach a particular Monthly Energy Production you'll need to know your locations Wind Zone, and Tower Height. From this information, the manufacturer can calculate your Energy production, for a given Power Rated Wind Turbine.

In easy terms, once you decide the general type, or model of your Wind Turbine, contact the manufacturer with your location's Wind Zone, and proposed Tower Height. Manufacturers have sophisticated software with Wind Turbine Testing data to give you an accurate Monthly Energy Electricity Production. Contacting the manufacturer may seem an extra step, but its the best way to get the most accurate information.

If you know the Energy you want to produce, or the Power Rating of the Wind Turbines you wish to use, please refer to the **Quick Guide** for System Examples.

Chapter Three: Skystream 3.7 Wind Turbine for Grid Connection and Selling Electricity

The Skystream 3.7 is a Wind Turbine designed for home, and commercial applications. The Skystream can be used as a single wind turbine, or "bundled" into a mini Wind Farm. The limit to how many you can connect to the Grid will depend on your Electrical Service.

The power of the Skystream, as for all wind turbines, depends on the Wind Speed considered. The "power" rating of a wind turbine is useful information, but the "Energy" rating is what counts.

The Skystream is the most popular Home Grid Tie Wind Turbine in the country, with many diverse locations. As Qualified hardware, the Skystream has a long and proven track record. Most Utilities are familiar with Connecting Skystream Wind Turbines to their Electrical Service Grid. And, most Utilities have a program in place to offer Net-Metering using the Skystream.

The "energy" is what your Wind Turbine exports to the grid in a given time frame, usually monthly. A given wind turbine will have different "energy"production at different heights. Therefore, your tower choice will directly effect your Wind Farm's energy production, irregardless of wind turbine "power" ratings.

The Skystream 3.7 Wind Turbine:

From a Power point of view the Skystream 3.7 is rated at 2.1 Kilowatts, at a wind speed of 24.6 mph (11 m/s). The Nominal Power rating is 2.4 Kw (Kilowatts) in a wind speed of 29 mph (13 m/s).

Power ratings, however, don't tell you how much "energy" in (kWh) your wind turbine will produce on site - in actual operation. Energy production depends on your location's (wind resource), and the height of your tower. Actual energy production will also be effected by local obstacles. Select your Tower Site for elevation and clear of all buildings, trees, etc.

As an example, with a 70 foot tower, in a Wind Zone (III) location, a Skystream will produce an average of 500 kWh of Energy per Month. If you're a home owner with a monthly energy consumption over 500 kWh, the Skystream is appropriate for choice of Wind Turbine.

For additional energy output combine multiple Skystreams for your Wind power system for farms, ranches, and commercial power supplies. To size your Wind energy facility for Home Power check your electric bill for your total Monthly kWh consumption. Size your wind turbines, by energy, to be just below your Monthly kWh consumption, for best results.

Skystream 3.7 Wind Turbine Physical Characteristics:

Weight of Skystream (without Tower): 170 lb (77 Kg)
Diameter of Blades: 12 feet (3.72 m)
Operating Temperature Range: -40F to 122F (-40C to 50C)
Tower Heights: 44 Feet to 70 Feet

Wind Speed Characteristics:

Cut in Wind Speed: 6.7 mph (3 m/s)

Survival Speed: 140 mph (63 m/s)
Peak Power Wind Speed: 25 mph

Electrical Characteristics:

Skystream 3.7 Wind Turbines can be connected to the Grid in different formats, these include:

120/240 VAC, 60 Hz, 2-Phase (Split Single Phase), and 120/208, 60 Hz, 3 Phase

Your Electric Service Provider, (ask your Utility) will dictate which Version of the Skystream you'll use in your project. Grid Interconnection only occurs with Utility approval.

The Following Skystream Single Wind Turbine system is based on a Wind Zone III location, and 70 Foot Tower Height.

The following Sample System is for one (1) Skystream 3.7 for home use.

Example A: Energy Production 500 kWh per Month

Skystream 3.7 Wind Turbine System: 500 kWh per Month Average output (will vary with individual sites, wind zones, and tower heights).

Power Rating 2.1 Kw.

Parts List:

One (1) Skystream 3.7 Grid Tie specify 120/240 2-Phase (Split phase), or 240/440 3 Phase. One (1) Monopole Tower Kit Choose 44 Foot, or 70 Foot Monopole Tower Height One (1) Foundation Hinge Plate Kit. One (1) Foundation Installation Kit. One (1) Gin Pole Kit

Electrical wiring/Junction Box/Safety Disconnects/Fusing Site Specific. Site a single Skystream within 100 feet of your Home Junction Box for best results.

Do not exceed 400 feet from Home Junction Box. Be sure your Tower cannot Fall Directly on any Building in the event of a catastrophic event.

The Height of your Tower, will be the Radius, from the base, you wish to have clear at your site. Note: exceptions depend on local conditions.

Recommend 70 Foot Tower for maximum energy production.

Chapter Four: Wind Farms based on the Skystream 3.7 Wind Turbine for Home Power and Commercial Use

In this chapter we'll look at the combining Skystreams to produce your own Wind Farm. The key is in the "electrical" interconnection. In addition, when you place your wind turbines, there are limits to how tightly you can "pack" your wind turbines for maximum energy production.

Ideally, you'll have some information about your Wind Sites wind energy profile. If your Utility, or

State Energy Office has done surveys, they may have produced a "Wind Rose" diagram for your location, or somewhere nearby.

A "Wind Rose" diagram, gives you a view of which directions, and for how long, the wind blows across your site. Knowing the Prevailing wind direction gives you an idea how to orient your turbines relative to each other.

If the prevailing wind, say an On-shore wind, is coming from the West, for example. Then, orient your wind turbines in a line North to South.

Obstacles in the wind path produce "shadows" of wind turbulence. In our example, if you oriented your turbines West to East, then the wind "shadow" will lower the performance of the other turbines.

Back Wash is the term for wind effects after encountering an obstacle, you want to minimize this effect.

Ideally, wind turbine should be spaced together no closer than the height of their tower. Ideally, you'll want as much space between towers as your site allows. However, the trade-off is increased cost in interconnection wire. You'll need to calculate a few scenarios in your wind Site Plan.

System Example B: Energy Production 1,000 kWh per Month

Two (2) Skystream 3.7 Wind Turbines System: 1,000 kWh per Month Average output (will vary with individual sites, wind zones, and tower heights). Power Rating 4.2 Kw.

Parts List:

Two (2) Skystream 3.7 Grid Tie specify 120/240 2-Phase (Split phase), or 240/440 3 Phase.

Two (2) Monopole Tower Kit Choose 44 Foot, or 70 Foot Monopole Tower Height

Two (2) Foundation Hinge Plates

Two (2) Foundation Kits

One (1) Gin Pole Kit

Electrical wiring/Junction Box/Safety Disconnects/ Fusing Site Specific.

Example C: Energy Production 1,500 kWh per Month

Three (3) Skystream 3.7 Wind Turbines System: 1,000 kWh per Month Average output (will vary with individual sites, wind zones, and tower heights). Power Rating 6.3 Kw.

Parts List:

Three (3) Skystream 3.7 Grid Tie specify 120/240 2-Phase (Split phase), or 240/440 3 Phase.

Three (3) Monopole Tower Kit Choose 44 Foot, or 70 Foot Monopole Tower Height

Three (3) Foundation Hinge Plates

Three (3) Foundation Kits

One (1) Gin Pole Kit

Electrical wiring/Junction Box/Safety Disconnects/Fusing Site Specific.

Example D: Energy Production 2,000 kWh per Month

Four (4) Skystream 3.7 Wind Turbines System: 1,000 kWh per Month Average output (will vary with individual sites, wind zones, and tower heights). Power Rating 8.4 Kw.

Parts List:

Four (4) Skystream 3.7 Grid Tie specify 120/240 2-Phase (Split phase), or 240/440 3 Phase.

Four (4) Monopole Tower Kit Choose 44 Foot, or 70 Foot Monopole Tower Height

Four (4) Foundation Hinge Plates

One (1) Gin Pole Kit

Electrical wiring/Junction Box/Safety Disconnects/Fusing Site Specific. For residential interconnection check with Electric Service Provider for 200 Amp service.

Example E: Energy Production 2,500 kWh per Month

Five (5) Skystream 3.7 Wind Turbines System: 3,000 kWh per Month Average output (will vary with individual sites, wind zones, and tower heights). Power Rating 10.5 Kw.

Parts List:

Four (5) Skystream 3.7 Grid Tie specify 120/240 2-Phase (Split phase), or 240/440 3 Phase.

Four (5) Monopole Tower Kit Choose 44 Foot, or 70 Foot Monopole Tower Height

Four (5) Foundation Hinge Plates

One (1) Gin Pole Kit

Electrical wiring/Junction Box/Safety Disconnects/
Fusing Site Specific. Be sure to space your Wind
Turbines as far apart as you can, yet close enough to
your interconnection transformer so not to have
extreme distances.

Each wind turbine may have differing outputs, at
any given moment. The Power conditioning
protocols, and controls, balance each turbines
output with the Grid power phase, sensed, and
monitored constantly by the power controller.

Chapter Five - The 442SR Wind Turbine for Grid Connection and Selling Electricity

This chapter examines the Xzeres 442SR Wind Turbine rated at 10,000 watts. The X-442 is a robust wind turbine engineered to be a work-horse. Recent innovations in power conditioning provide excellent response (electricity production) over a wide range of wind speeds. Wide ranges of response give you a greater energy production which is the "gold" of wind harvesting.

Internal "power control circuits" give the X-442 a low cut-in speed, bringing the turbine into production is light winds. The sweet spot for the X-442 begins in wind speeds over 14 mph. As power density increases, exponentially, with increasing wind speed, the X-442 cranks in the middle wind speeds, from 14-mph to rated wind speeds (28 mph).

The 442 is a large wind turbine, weighs 2,300 lbs which sits atop your Tower.

Your Tower must be robust, and engineered for proper ratings. The manufacturer has anticipated these requirements, and has pre-engineered, and PE stamped plans ready for your use.

Producing 10,000 watts of power, at peak, the 442 has an impressive output at nearly 5 watts per pound.

Your 10,000 Watt Wind Turbine will need to mounted on a Tower. Your choices are Lattice type, and Monopole Type. Your Foundation comes in two types: Mat, and Pedestal type, respectively.

Your Wind Turbine power system will include the 442 Wind Turbine (Masthead) Assembly, Tower and Foundation, Interconnection Wiring, Main Power Panel, or Junction Boxes if specified by your local NEC code.

The 442s mount on Towers from 44 feet to 120 feet high. Assembly of Tower, and Masthead Wind Turbine, is usually done on the ground. Crane needed if not lifting by Gin Pole, and Ground Mounted Winch.

Xzeres 442SR Wind Turbine Physical Characteristics:

Weight: 442SR Wind Turbine (without Tower): 2,300 lb (1,043 Kg)
Diameter of Blades: 23.6 feet (7.2 m2)
Blade Swept Area: 442 square feet (41 m2)
Operating Temperature Range: -40 F. to 122F (-40 C. to 50 C.)

Wind Speed Characteristics:

Cut in Wind Speed: 5 mph (2.2 m/s)
Survival Speed: 140 mph (63 m/s) gust

Electrical Characteristics:

The 442SR Wind Turbine can be connected to the Grid in different formats, these include:

240 VAC, 60 Hz, 2-Phase (Split Single Phase), and 240/440, 60 Hz, 3 Phase

Your Electric Service Provider, (ask your Utility) will dictate which Version of the 442SR you'll use in your project. Grid Interconnection only occurs with Utility approval.

Wind Farm Applications using the 442SR Wind Turbine:

The 442SR Wind Turbine produces about 2,500 kWh of electricity each Month (Wind Zone III, at 120 feet) as an average figure. If your Electric Bill for your Home, Office, Farm, Ranch, or commercial facility shows a consumption over 2,500 kWhs of energy, the 442SR can be a great solution.

Check your electric bill, or your facilities Monthly Energy Consumption in kWh per Month to size your Wind Farm.

The 442SR can be interconnected to the Grid, through your electric company (Electric Service Provider). The number of 442SRs you can install at your site only limited by your Electrical Service Hardware available at your site for Grid Connection.

Commercial loads are often 3 Phase, and have that service box available, onsite. The 442SR is engineered to interface with your existing electrical service. Qualified installers, can tap into your existing Utility Service interconnecting through your Sub-station, or Electrical Service Feed. The 442SR Wind Turbines are, essentially, plan, procure, prepare, install, commission, and Plug-N-Play.

Used in remote Rural settings, small Wind Farms can be constructed using the 442SR Wind Turbine. Using the same model of wind turbine, throughout your facility, keeps Planned Maintenance Schedule

(MSP) manageable, and replacement parts standardized.

Largest Wire size #8 AWG (10 mm2) for the 442SR Wind Turbine 240 VAC. Use 10 Amp Circuit Breakers in your breaker box.

For Electrical Loads, or Energy Production over 2,500 kWh per Month, the 442SR Wind Turbine is the most reliable, and well supported Wind Turbine in its class. Use the 442SR for US, and European voltage grid tie power production.

Example F: Energy Production 2,500 kWh per Month

One (1) 442SR Wind Turbine for 2,500 kWh per Month Average Electricity Production output (will vary with individual sites, wind zones, and tower heights). Power Rating 10 Kw.

Parts List:

One (1) 442SR Wind Turbine specify 240 2-Phase (Split phase), or 240/440 3 Phase.

One (1) Lattice, or Monopole Tower from or 70 to 120 Foot Tower Height

One (1) Foundation Kit

One (1) Mounting Plate Kit

Electrical wiring/Junction Box/Safety Disconnects/ Fusing Site Specific. Best to use 120 Foot Tower for maximum energy production. Lattice Tower offers best economic performance.

Example G: Energy Production 5,000 kWh per Month

Two (2) 442SR Wind Turbine, 120 Foot Tower Height for 5,000 kWh per Month Average output (will vary with individual sites, wind zones, and tower heights). Power Rating 20 Kw.

Parts List:

Two (2) 442SR Wind Turbine specify 240 2-Phase (Split phase), or 240/440 3 Phase.

Two (2) Lattice, or Monopole Tower from or 70 to 120 Foot Tower Height

Two (2) Foundation Kits

Two (2) Mounting Plate Kits

Electrical wiring/Junction Box/Safety Disconnects/ Fusing Site Specific.

Example H: Energy Production 7,500 kWh per Month

Three (3) 442SR Wind Turbines for 7,500 kWh per Month Average Electricity output (will vary with individual sites, wind zones, and tower heights). Power Rating 30 Kw.

Parts List:

Three (3) 442SR Wind Turbine specify 240 2-Phase (Split phase), or 240/440 3 Phase.

Three (3) Lattice, or Monopole Tower from or 70 to 120 Foot Tower Height

Three (3) Foundation Kits

Three (3) Mounting Plate Kits

Electrical wiring/Junction Box/Safety Disconnects/ Fusing Site Specific.

Example I: Energy Production 10,000 kWh per Month

Four (4) 442SR Wind Turbine for 10,000 kWh per Month Average output (will vary with individual sites, wind zones, and tower heights). Power Rating 40 Kw.

Parts List:

Four (4) 442SR Wind Turbine specify 240 2-Phase (Split phase), or 240/440 3 Phase.

Four (4) Lattice, or Monopole Tower from or 70 to 120 Foot Tower Height

Four (4) Foundation Kits

Four (4) Mounting Plate Kits

Electrical wiring/Junction Box/Safety Disconnects/ Fusing Site Specific.

Example J: Energy Production 20,000 kWh per Month

Eight (8) 442SR Wind Turbine for 20,000 kWh per Month Average output (will vary with individual sites, wind zones, and tower heights). Power Rating 80 Kw.

Parts List:

Eight (8) 442SR Wind Turbine specify 240 2-Phase (Split phase), or 240/440 3 Phase.

Eight (8) Lattice, or Monopole Tower from or 70 to 120 Foot Tower Height

Eight (8) Foundation Kits

Eight (8) Mounting Plate Kits

Electrical wiring/Junction Box/Safety Disconnects/
Fusing Site Specific.

Chapter Six - Installing, and Commissioning your Wind Farm Energy Facility

Wind turbines are usually installed at remote sites. However, as more home owners, and small business owners, face increasing and uncertain electricity costs, more are turning to harvesting and selling Wind Power.

Before installing your wind turbine, of course, is the planning phase. Is Wind Right for your Site? The short answer: if your site is "windy" then you have your first Green Light.

In this Book we're looking at the Skystream 3.7 wind turbine (2.1 Kw), and the 442SR wind turbine (10 Kw).

Installing your wind turbine, or wind turbines starts with choosing your site. In chapters above we've looked at site selection, and site access for pickup trucks, or cranes, as required. Tilt-up towers don't require Cranes.

Examine your Soil Type onsite, is it rocky soil, or clayish. Based on this information the Manufacturer will provide drawings which you can use to Pour your own pad, or have a contractor do it.

Planning your Wind Facility you'll be developing your Site Plan. This Wind Project description will include Wind Resource Data, Site Selection, Site Access, Soil Type, Tower Foundation, Wind Tower itself, Wind Turbine Assembly, Installation Plan, and Electrical Interconnection details.

Site Preparation and Foundations:

Foundations are engineered based on the size of the Wind Turbine, the height of the Tower, and the Site's Soil-Type. Towers are engineered for the Lateral Thrusts, and other "loads" endured by the structure under normal use, and extreme wind events. Your site preparation will be dictated by your local site conditions, and your soil type, as defined by ANS/TIA-222G standards, for your foundation engineering.

Fortunately, your Wind Turbine Manufacturer (Tower Manufacturer) has done this for a variety of soil types, and will provide drawings which are PE stamped, and ready to use.

Structural Backfill material should be compacted in 10" loose lifts, within 98% of the maximum dry density of your soil. Depending on whether your Tower Foundation will be Mat Type, or Pedestal Type determines the amount of material you should prepare. Mat-type foundations require more concrete, but take less labor.

Use a Mat-type foundation if your site is in a mild climate.

If your site is in an extreme location, such as high altitude, extreme temperatures, or Monsoons, then use a Pedestal type foundation. The Pedestal is more labor intensive, but uses less concrete. The Pedestal type has more concrete "buried" than is visible at the surface. This underground "pyramid" shape is very well based. If you're pouring your own foundation, see Owners Manual for additional specifications on Rebar, and aggregate required.

If your Wind Turbine is some distance from your Electrical Interconnect, a very common case, then trench the ground for your wires to protect, and isolate them. Trenching, will be site specific, but will be included in your Site Preparation.

Prepare Grounding for your Wind Turbines (grounding rods etc.) Specific grounding specifications, if you're installing the Wind Turbine yourself, are listed in the Installer's Manual depending on the Tower you're using. General reference for Grounding see section 250 National Electric Code NEC ANSI/NFPA 70. NEC (USA), and IEC 60364-5-54.

Site preparation and Foundation pouring will provide a Level, stable grade to install your Mounting Plates to which your Tower Base will mount. Lattice Towers will Bolt to the Mounting Plate. Tilt-up Wind Towers with have a Hinge Plate to "Tilt Up" your Tower with a winch.

Note: Remove all loose material from the bottom of your excavation for your Foundation before pouring concrete.

Tower Assembly:

Be sure your site provides room to layout your tower for assembly on the ground with the bottom of the tower near the Base. Wind Towers, either Monopole, or Lattice type, come in sections which must be assembled onsite.

Assemble your Tower starting with the Top sections, of the Tower, and Work your way to the Bottom.

Next, assemble your Wind Turbine Assembly (Wind Turbine), near the top of the tower because you'll preassemble everything before your lift, and attach (Bolt) the Assembly to the "Top" of the tower - while still on the ground.

Connect your wires to the Wind Turbine Assembly (unconnected to anything on the other end, and taped off), and "string" your wires down the tower (if Lattice) or Through the Tower, if Monopole to the Output Junction Box located at the tower base.

Your two choices of "How" to lift, or erect your tower, after assembly on the ground, is a Crane, for direct lift, or Tilt-Up which is Ground based Winching, with a Gin pole for leverage.

In all cases you'll build your tower, and connect your wind turbine while on the ground. Note: When you connect your Blades to your Wind Turbine (Last Step), you can lift the tower on one end up 6 feet, or so, so when you Connect your Turbine Blades to the Nose Hub, they do not contact the ground.

Electrical Interconnection:

The Output Junction box is at the bottom of your Tower. After you Lift your Tower, you'll interconnect your Output Junction Box with wiring which will lead to your Electric Service Provider Transformer, or other Service Panels. (Utility will specify).

Some local electric codes require a Safety Disconnect Box, the manufacturer will tell you which, and/or Junction Box Fusing. Install Disconnects between the Wind Turbine, and your Electrical Wire connection to your Electric Utility.

Permitting and Commissioning:

Wind Turbines used in Large Wind Farms face a multitude of permitting issues (Regulatory Affairs). Formally, permitting is on the Local, State, and Federal level, if you're using Federal lands.

Smaller wind turbines, such as those included in this eBook, have an easier process because many Utilities are now experienced with Small Wind Farm facilities. Rebate programs have been created to expedite your Wind Energy Facility when connecting to a grid.

Under most Utility Net Metering projects your site will be inspected shortly after you file your Site Plan with the utility, and/or state regulatory body. After, the approval process, your site will be authorized for your installation. Once your Wind Turbines are installed, and interconnected, then a Site Inspection will be scheduled to confirm your Wind facility.

Upon this inspection your Wind Turbine facility is officially "commissioned." Now, you're eligible for all applicable Rebates, Tax Credits, Grants, and Incentives promoted in your State. Other than Federal programs, State programs range greatly.

Your Utility will have the information you need to go through the permitting process, relative to the local codes, which you must follow. To begin, prepare your Site Plan, and your Impact Study.

Chapter Seven - Rebates, Tax Credits, Grants, and Financial Incentives for your Wind Farm

Wind turbines enjoy an amount of of public support which can rebate as much as 85% of your installed cost, depending your Sites Qualifications.

Your Wind Farm will be producing Carbon-free electricity saving 2.2 lb. of Carbon Dioxide per kwh of energy. For example, a production of one Skystream Wind Turbine, producing 500 kWh of clean energy per month displaces over 1/2 Ton of Carbon Dioxide, per month, being spewed into the environment.

Because of this social value, there has been an emergence in the last decades of a Carbon Market, and other Financial mechanisms to monetize and reward Wind Power technology - in its use.

Depending on your Utility, and the Rules of your State, and with precedence Federally, the "value" of carbon free energy ranges from 1/2 cent to 2.2 cents Per kWh of electricity. The great advantage of these emerging Carbon markets allow you to "sell" All of your projects Carbon credits in one deal. It sets the price you're paid, but "leveraging" or "selling" your 20 years worth of carbon credits is another value you can tap.

Large Wind Farms sell their carbon tax credits, or Production Tax Credits (PTC) all at once, and apply that capital to their initial capital cost.

The Skystream 3.7, and 442SR Wind Turbines, are pre-qualified for Tax Credits, Rebates, Grants, and other incentives.

Federally, Clean Energy projects, such as your Wind Farm qualify for a 35% Tax Credit, at the time of this writing. This is a Tax-Credit, so you need to have a Tax Liability, to apply it to, however, the IRS will allow you to apply it over several years. Check with your CPA for exact periods.

The United States Department of Agriculture has Rural Development Grants for Wind Projects. Qualifying Facilities (QF) are in designated Rural Zones. Click this link for more information. USDA grants up to 25% of your installed costs.

All of these grants, rebates, and programs Only apply After you've installed, and commissioned your Wind Turbine. Therefore, these funds can be "refunded" to you after your Wind Farm is up and operational. However, the time frame is not too unreasonable, usually within 6 month to one year after operational status depending on program.

Few capital investments, in the public market, exist which are as strong as Wind Farms for building a "back-end" of support to compensate for expensive equipment on the front end. Public financial support for your Wind Farm is warranted, and you can tap into this funding as you tap into the wind.

As qualified equipment, and with greater experience by the Utilities, Wind Turbines, including Skystream, and 442SR, are ideal for Residential, Commercial, and Remote Site power production for Export to the electrical Grid.

Chapter Eight - Wind Farm Opportunities

Wind Farms are power plants. Wind Farms, however, are different operationally from Power Plants that burn fuels, such as coal, oil, and natural gas, with steady outputs. Wind Turbines produce variable energy. Fortunately, today's Power Conditioning equipment is advanced in "matching" the electrical production of your Wind Turbine with the grid.

Over a wide range of wind conditions your Grid Connected Wind Facility, of one or multiple wind turbines, can produce energy "over time" from which you get paid.

Add financial "incentives," and you see why Wind Farms have been the fastest growing Power Plant segment worldwide. From a business stand-point, Wind Farms have strong earnings potential. Proper turbine selection, siting, installation, and planned maintenance makes wind turbines profitable in many markets.

The greatest financial advantage? No fuel costs, and "back end" income from environmental offsets. Clean energy has value, and wind turbines deliver strong advantages, especially at remote sites, where fuels are costly to deliver.

Wind Farms can be used as a power supply for Homes, Businesses, and Remote Sites selling power back to the utility against your bill. Water Pumping, and Remote Communication power supplies often use Wind Turbines because they work with minimal maintenance. This eBook is focusing on Grid Tied wind power systems for grid connected energy production.

Wind power has been very successful around the world because Wind Turbine technology has matured. After decades of field experience, investors, who have horror stories from the 70s, recover in the 80s, and began to prosper in the 90s. In this decade, wind turbine manufacturer offer a reliable, and mechanically supported range of Wind Turbines which make the grade.

Financially, the two opportunities for selling electricity to Utilities is through either Net-Metering, or Power Purchase Agreements (PPA).

To qualify for Net Metering you need to be connected to a Utility (Electric Service Provider) which supports net metering. Second, you need to size your facility energy production Just-Below your average Monthly Energy consumption.

Power Purchase Agreements are entirely different, and do not depend on your current consumption, at a particular location. Utilities, in recent decades, have divested their ownership of Power Plant facilities, and opt now to outsource production. This is your Wind Opportunity for Wind Farm development.

The contract between the Independent Power Producer (IPP), and the Utility is the Power Purchase Agreement (PPA). There is no limit to the size, or amount of Kilowatt-hours (kWh) which you produce at your facility. The market is driven by the "Demand" of a given utility.

Alaska, for example, has tremendous Wind Power potential because remote villages are almost exclusively powered with Diesel Generators. Expensive, and environmentally dangerous, transporting Diesel fuel through pristine areas to remote villages is problematic.

Wind Farm opportunities range from Home owners using Net Metering, to Independent Power Producers supplying electricity to Utilities.

Remote location electrification, and home power offer the vast opportunities for wind turbine applications for producing electricity to sell to Utilities large and small.

Chapter Nine: Quick-Guide to Wind Farm System Examples from 500 to 20,000 kWh per Month

Listed below are Wind Energy Systems, formally called Wind Energy Conversion Systems (WECS), are known popularly as "Wind Farms." The following example systems, below, are listed by Monthly Energy Production:

Example A: Skystream 500 kWh Energy per Month

Example B: Skystream 1,000 kWh Energy per Month

Example C: Skystream 1,500 kWh Energy per Month

Example D: Skystream 2,000 kWh Energy per Month

Example E: Skystream 2,500 kWh Energy per Month

Example F: Xzeres 442SR 2,500 kWh Energy per Month

Example G: Xzeres 442SR 5,000 kWh Energy per Month

Example H: Xzeres 442SR 7,500 kWh Energy per Month

Example I: Xzeres 442SR 10,000 kWh Energy per Month

Example J: Xzeres 442SR 20,000 kWh Energy per Month

Note: Often Rebate programs define a Wind Turbine by Power Rating to calculate rebates. Use the following list to find the Power Rating you're looking for in Wind Turbine Energy Systems.

Wind Energy Systems by Listed by Power Rating:

System A: 2.1 Kilowatts, (2,100 watts)

System B: 4.2 Kilowatts, (4,200 watts)

System C: 6.3 Kilowatts, (6,300 watts)

System D: 8.4 Kilowatts, (8,400 watts)

System E: 10.5 Kilowatts, (10,500 watts)

System F: 10 Kilowatts, (10,000 watts)

System G: 20 Kilowatts, (20,000 watts)

System H: 30 Kilowatts, (30,000 watts)

System I: 40 Kilowatts, (40,000 watts)

System J: 80 Kilowatts, (80,000 watts)

Convert the wind energy blowing across your property into cash. Wind Turbines offer unique opportunities in clean energy production, with strong performance in the field, and strong economic performance on the balance sheet.

Wind Zone Descriptions Wind Speeds, and Power Density.

Leveraging your Capital Costs with public incentives can lower your system costs, as much as 85% if your Wind project is a Qualifying Facility (QF) relative to the programs available. To be (QF) with your project depends on your site specific situation, and, you have Utility, Local, State, and Federal incentives you can tap into to lower your costs dramatically.

Before you begin your Wind Farm project, check with your Utility, and your State Energy office for the roadmap they've prepared for Electricity Consumers, to become Electricity Producers.

For more information on Wind Turbines, Wind Farms, and other clean energy topics, please visit **Solardyne.com** on the worldwide web.